Nature School

LET'S LOOK FOR WORMS!

By Seth Lynch

Please visit our website, www.garethstevens.com. For a free color catalog of all our high-quality books, call toll free 1-800-542-2595 or fax 1-877-542-2596.

Library of Congress Cataloging-in-Publication Data

Names: Lynch, Seth, author.
Title: Let's look for worms! / Seth Lynch.
Description: Buffalo, New York : Gareth Stevens Publishing, [2024] |
  Series: Nature school | Includes index. | Audience: Grades K-1
Identifiers: LCCN 2022045123 (print) | LCCN 2022045124 (ebook) | ISBN
  9781538286326 (library binding) | ISBN 9781538286319 (paperback) | ISBN
  9781538286333 (ebook)
Subjects: LCSH: Worms–Juvenile literature.
Classification: LCC QL386.6 .L96 2024 (print) | LCC QL386.6 (ebook) | DDC
  592/.3–dc23/eng/20221013
LC record available at https://lccn.loc.gov/2022045123
LC ebook record available at https://lccn.loc.gov/2022045124

First Edition

Published in 2024 by
Gareth Stevens Publishing
2544 Clinton St,
Buffalo, NY 14224

Copyright © 2024 Gareth Stevens Publishing

Editor: Kristen Nelson
Designer: Claire Wrazin

Photo credits: Cover BERNATSKII IURII/Shutterstock.com; p. 1 Andrei Metelev/Shutterstock.com; p. 5 Elizabeth A.Cummings/Shutterstock.com; pp. 7, 19 galitsin/Shutterstock.com; pp. 9, 24 (annuli) hsagencia/Shutterstock.com; pp. 11, 23 Rawpixel.com/Shutterstock.com; pp. 13, 24 (burrow, soil) E-lona/Shutterstock.com; p. 15 Ksenia Shestakova/Shutterstock.com; p. 17 NatalyWatson/Shutterstock.com; p. 21 Liz Weber/Shutterstock.com

All rights reserved. No part of this book may be reproduced in any form without permission in writing from the publisher, except by a reviewer.

Printed in the United States of America

CPSIA compliance information: Batch #CS24GS: For further information contact Gareth Stevens, New York, New York at 1-800-542-2595.

Contents

Wiggly Worms 4

Worm Watch 10

Good for the Garden . . . 16

Night Crawlers 20

Words to Know 24

Index 24

It rained today.
We see a worm!

It is a reddish gray color.
It is shaped like a tube!

It's body is made up of rings.
These parts are annuli.

It is sunny today.
Where are the worms?

Worms live in soil.
They burrow in it.

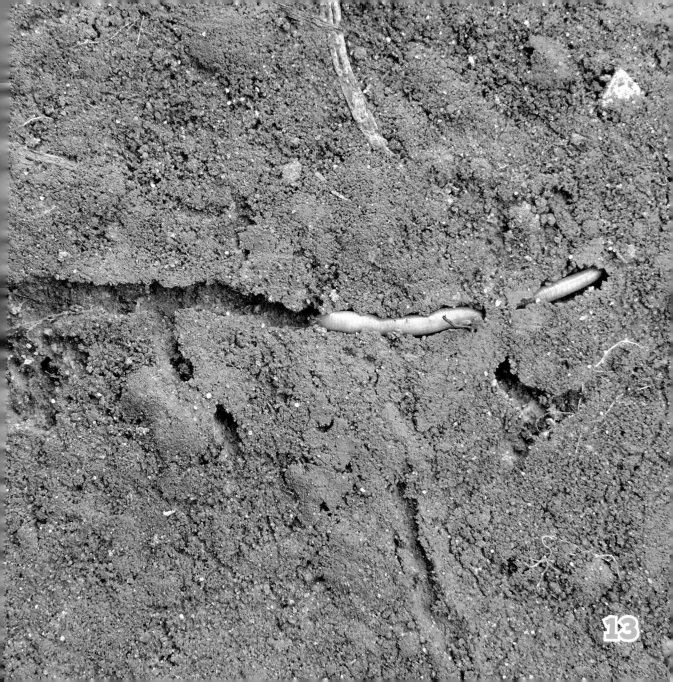

We dig in the garden.
We find a worm!

Worms help gardens grow. Their burrows bring air into the soil.

They eat dead plants. Their waste makes good soil.

Worms come out at night.
They are called night crawlers.

Let's look for more worms!

Words to Know

annuli   burrow   soil

Index

body, 6, 8   garden, 14, 16
color, 6   soil, 12, 16, 18